Tbilisi Mathematical Journal

Volume 3 (2010)

ISBN 978-1-84890-035-6

College Publications
Scientific Director: Dov Gabbay
Managing Director: Jane Spurr
Department of Computer Science
King's College London, Strand, London WC2R 2LS, UK

http://www.collegepublications.co.uk

Original cover design by Maedium, Utrecht.
Adapted and produced by Laraine Welch
Printed by Lightning Source, Milton Keynes, UK

Tbilisi Mathematical Journal

VOLUME 3 (2010)

Editor-in-Chief. Hvedri Inassaridze (Tbilisi)

Managing Editors.
David Applebaum (Sheffield), Benedikt Löwe (Amsterdam),
Russell Lyons (Bloomington IN)

Editors.

Malkhaz Bakuradze (Tbilisi)	John Ball (Oxford)
Peter Bates (East Lansing MI)	Guram Bezhanishvili (Las Cruces NM)
Victor Buchstaber (Moscow)	Dan Burghelea (Columbus OH)
Maria Carro (Barcelona)	Alberto Cattaneo (Zürich)
Christophe Cheverry (Rennes)	Alex Chigogidze (Greensboro NC)
Karen Collins (Middletown CT)	Wojciech Gajda (Poznań)
Ross Geoghegan (Binghamton NY)	Amiran Gogatishvili (Prague)
Marco Grandis (Genova)	Joseph Gubeladze (San Francisco CA)
Graham Hall (Aberdeen)	Nick Inassaridze (Tbilisi)
George Janelidze (Cape Town)	David Jordan (Sheffield)
George Khimshiashvili (Tbilisi)	Manfred Kolster (West Hamilton ON)
Mark Kon (Boston MA)	Pekka Koskela (Jyväskylä)
Michael Lacey (Atlanta GA)	Kirill Mackenzie (Sheffield)
Kohji Matsumoto (Nagoya)	Jan van Mill (Amsterdam)
Vicente Muñoz (Madrid)	David Natroshvili (Tbilisi)
Ryszard Nest (Copenhagen)	Lars-Erik Persson (Luleå)
Teimuraz Pirashvili (Leicester)	Roger Plymen (Manchester)
Tomas Recio (Santander)	Gerhard Röhrle (Bochum)
Eugene Shargorodsky (London)	Theodore Simos (Tripolis)
Rainer Vogt (Osnabrück)	Efim Zelmanov (La Jolla CA)

Tbilisi Mathematical Journal is an electronic journal covering all of mathematics. Authors are invited to submit high-quality research articles via the EasyChair webpage of the TMJ.

http://www.tcms.org.ge/Journals/TMJ/

Tbilisi Mathematical Journal
VOLUME 3 (2010)

The divisor class group of a Quot scheme

Rafael Hernández and Daniel Ortega*

Departamento de Matemáticas, Universidad Autónoma de Madrid, Ciudad Universitaria de Cantoblanco, 28049 Madrid, Spain

E-mail: {rafael.hernandez, daniel.ortega}@uam.es

Abstract

The divisor class group of the Quot scheme parameterizing quotients of fixed degree and rank of a trivial bundle on a curve is computed. The method used is a reduction to the case of rank 0 quotients.

2000 Mathematics Subject Classification. **14H60**.

Keywords. Vector bundle, Quot scheme, divisor class group.

1 Introduction

In his paper [14], Strømme computed the Chow ring of the Quot scheme parameterizing quotients, of rank r and degree d, of a trivial bundle of rank n on $\mathbb{P}^1$. For the Chow group of codimension one cycles, Strømme obtained $\mathbb{Z}$ if $r = n - 1$ and $\mathbb{Z} \oplus \mathbb{Z}$ for $r < n - 1$.

The aim of this note is to generalize a small part of Strømme's work to curves of arbitrary genus. Specifically, we want to compute the Chow group of codimension one cycles. This Chow group coincides with the divisor class group of the scheme. Let us define

$$b_1(g,r,n) := \begin{cases} 2g & \text{for } n - r = 1 \\ 1 + \frac{n^{n-r-1}}{(n-r-1)!}\left((2g-1) - (g-1)\sum_{k=1}^{n-r-1} \frac{k!}{n^k} \right) & \text{for } n - r \geq 2 \end{cases}$$

$$b_2(g,r,n) := (n-r)(2+n^2)g + 2(n-r)$$

for integers g, r and n verifying $g \geq 1$, $0 < r < n$.

Theorem 1.1. Let C be an algebraic curve of genus $g \geq 1$, over an algebraically closed field of characteristic zero $\mathbb{K}$, and let $Q(n,r,d)$ be the Quot scheme parameterizing degree d, rank r quotients of a trivial bundle of rank n on C. Let us denote by $\mathrm{Cl}(Q(n,r,d))$ the divisor class group of the Quot scheme. For $d > \max\{b_1(g,r,n), b_2(g,r,n)\}$ we have:

1. if $r = n - 1$, then $\mathrm{Cl}(Q(n,r,d)) = \mathbb{Z} \oplus \mathrm{Pic}(\mathrm{Jac}^d(C))$;

---

*The second author was partially supported through grant MTM2007-63582 of the *Ministerio de Educación y Ciencia*, Spain. We thank several referees of a previous version of this paper for many useful suggestions that, we believe, have improved our presentation.

Tbilisi Mathematical Journal 3 (2010), pp. 1–15.

Tbilisi Centre for Mathematical Sciences & College Publications.

Received by the editors: 14 October 2009; 31 May 2010.

Accepted for publication: 23 July 2010.

    2. if $r < n - 1$, then $\mathrm{Cl}(Q(n, r, d)) = \mathbb{Z} \oplus \mathbb{Z} \oplus \mathrm{Pic}(\mathrm{Jac}^d(C))$.

    In the first case, $Q(n, n - 1, d)$ is a projective bundle on the Jacobian of C [2, Corollary 4.23], and then (1) holds. We assume, from now on, that $r < n - 1$ and, as a consequence, that $n \geq 3$.

    On non-rational curves the Quot scheme, $Q(n, r, d)$, is, for d big enough with respect to $n, r,$ and g, irreducible and generically reduced, but it is neither smooth nor reduced. However, we can define its divisor class group, $\mathrm{Cl}(Q(n, r, d))$, as the divisor class group of the reduced structure.

    The divisor class group, or the Picard group, of a scheme is an interesting object that has been computed in a number of cases, three of them related to our work [6, 5, 11].

    Moreover, Quot schemes, classically studied in order to construct moduli spaces of vector bundles, have been recently the object of new interest, mainly because of their relation to the Kontsevich moduli space of stable maps [12].

    The tools we use are, basically, [7, Proposition 1.8 & Theorem 3.3] which we can apply because the divisor class group of the reduced scheme is the same group as the Chow group of divisors.

    The second section of the paper is a short review of Quot schemes, and then, in the third section, we reduce the computation from $Q(n, r, d)$ to some open sets in $Q(n, r, d)$, until we arrive at one whose Picard group we can compute. These reductions are based on work of Bertram, Kirwan, and Bifet [1, 10, 4].

2   The Quot scheme

We begin by recalling, in the particular situation needed for our result, the basic properties of the Quot scheme.

    Let C be an algebraic curve of arbitrary genus g, over an algebraically closed field $\mathbb{K}$ of characteristic zero. Let $\mathbb{K}^n$ be a $\mathbb{K}$-vector space of dimension n, and let us denote by $\mathcal{O}_C^n$ the trivial vector bundle of rank n on C, with fibres isomorphic to $\mathbb{K}^n$. Fix the polynomial $P(t) = r(t + 1 - g) + d$ and consider the contravariant quotient functor that assigns to each $\mathbb{K}$–scheme, X, the set of isomorphism classes of quotients $\mathcal{O}_{C \times X}^n \to \mathcal{F} \to 0$, such that $\mathcal{F}$ is flat over X with relative Hilbert polynomial $P(t)$. Two quotients, $\mathcal{O}_{C \times X}^n \xrightarrow{f} \mathcal{F} \to 0$ and $\mathcal{O}_{C \times X}^n \xrightarrow{g} \mathcal{F}' \to 0$, are said to be isomorphic if there is an isomorphism $\varphi : \mathcal{F} \to \mathcal{F}'$ such that $\varphi \circ f = g$.

    This quotient functor is representable by a projective $\mathbb{K}$-scheme ([15, 1.5] or [8]), the Grothendieck Quot scheme, $Q(n, r, d)$, having a universal property: families of quotients, as above, are functorially in one to one correspondence with morphisms $X \xrightarrow{f} Q(n, r, d)$ of $\mathbb{K}$–schemes. The universal property is associated to a universal family of quotients $\mathcal{O}_{C \times Q(n,r,d)}^n \to \mathcal{E} \to 0$, such that a family of quotients, defined on $C \times X$, is the pullback by $1_C \times f$ of the universal family.

In particular, if $q \in Q(n, r, d)$ is a point, the restriction of the universal family to $C \times \{q\}$ is the quotient of $\mathcal{O}_C^n$ parameterized by q. We will denote by

$$0 \to N_q \to \mathcal{O}_C^n \to E_q \to 0$$

the exact sequence obtained from the quotient parameterized by q.

The group $\mathrm{GL}(n)$ acts on $Q(n, r, d)$ because each $g \in \mathrm{GL}(n)$ is an automorphism of $\mathcal{O}_C^n$, and can be composed with a quotient, $\mathcal{O}^n \xrightarrow{f} E \to 0$, to give a new quotient $f \circ g$.

The tangent space to $Q(n, r, d)$ at a point q can be identified with $H^0(N_q^\vee \otimes E_q)$, and the obstruction space with $H^1(N_q^\vee \otimes E_q)$, a fact we will use later on.

We will denote by π the natural map from $Q(n, r, d)$ to $\mathrm{Jac}^d(C)$, sending the point q to the top exterior power of $N_q^\vee$.

The scheme $Q(n, r, d)$ is a natural compactification of the scheme $M(n, r, d)$, parameterizing morphisms of degree d from C to the Grassmannian $\mathrm{Gr}(r, n)$ of r-dimensional quotients of $\mathbb{K}^n$, because $M(n, r, d)$ is just the open subscheme parameterizing locally free quotients. The open set $M(n, r, d)$ is clearly invariant by the action of $\mathrm{GL}(n)$ on $Q(n, r, d)$.

If the genus of C is 0, the scheme $Q(n, r, d)$ is irreducible and smooth, and for higher genus it is known that, for d large enough, $Q(n, r, d)$ is irreducible and generically reduced [2, Theorem 4.28]; in the next lemma we will compute an explicit bound, and we follow, just for the proof of the lemma, the notations of that paper.

Then, $\overline{\mathcal{M}}_Q(d, n-r, n)$ is, as in [2], a compactification, defined as a Quot scheme, of the space of maps from the curve C to the Grassmann variety of linear subspaces of dimension $n - r$ in $\mathbb{K}^n$. We denote, in the rest of the paper, this same scheme by $Q(n, r, d)$.

Lemma 2.1. The scheme $\overline{\mathcal{M}}_Q(d, r, n)$, for $g, r \geq 1$, is irreducible and generically reduced of dimension $nd - r(n - r)(g - 1)$ whenever

$$d > f(g, r, n) := \begin{cases} 2g - 1 & \text{for } r = 1 \\ \frac{n^{r-1}}{(r-1)!}(2g - 1) - \left(\frac{n^{r-1}}{(r-1)!} \sum_{k=1}^{r-1} \frac{k!}{n^k} \right)(g - 1) & \text{for } r \geq 2 \end{cases}$$

Proof. Assume, according to [2, Theorem 4.28], that $f(g, r, n)$ is a function such that $\overline{\mathcal{M}}_Q(d, r, n)$ is irreducible and generically reduced of the expected dimension, $nd - r(n - r)(g - 1)$, for all $d \geq f(g, r, n)$. This function is not explicitly computed in [2], and its existence is proved inductively. Since every component of the Quot scheme always has, at least, the expected dimension $nd - r(n - r)(g - 1)$ (cf. [8]), the result follows if we find a lower bound for d such that any possible component of the complement $\overline{\mathcal{M}}_Q(d,r,n) \setminus \overline{\mathcal{M}}_Q^s(d,r,n)$ (cf. [2] for the notation) has dimension smaller than $nd - r(n-r)(g-1)$.

As part of the proof of [2, Theorem 4.28], it is shown that if $r > 0$ and $d \geq r(2g-2)$ are integers such that for all non–negative integers d_q, d_s, r_q, r_s satisfying the conditions

$$d_s + d_q = d, \quad r_s, r_q > 0, \quad r_s + r_q = r, \quad \text{and} \quad d_s r_q - d_q r_s > 0 \quad (2.1)$$

we have

$$n(f(g, r_q, n) - d_q) - (d_s r_q - d_q r_s) - r_s r_q (g-1) < 0, \quad (2.2)$$

then $Q(n, r, d)$ is irreducible and generically reduced of the expected dimension.

We are assuming here that we know the function $f(g, r_q, n)$ for $r_q < r$, and we need to find $f(g, r, n)$ such that for $d > f(g, r, n)$ and all the possible choices as in (2.1) the inequality (2.2) holds.[1]

Starting with the observation that for $r = 1$ and any n, $\overline{\mathcal{M}}_Q(d, 1, n)$ has these properties as soon as $d > 2g - 1$, we will compute bounds for all ranks r.

For fixed g and n, let us define $f_r := f(g, r, n)$. The sequence f_r has to satisfy, under the conditions (2.1), the inequality (2.2), i.e.,

$$n(f_{r_q} - d_q) - r_s r_q (g-1) + d_q r < d r_q. \quad (2.3)$$

For (2.3), it is sufficient (as $r < n$) to obtain the inequalities

$$\frac{n f_{r_q} - r_s r_q (g-1)}{r_q} < d. \quad (2.4)$$

We determine the largest fraction on the left hand side of (2.4). Let us denote by

$$F(m) := n \frac{f_m}{m} - (r - m)(g - 1)$$

one of these fractions. Then

$$F(m) - F(m-1) = n \left(\frac{m(f_m - f_{m-1}) - f_m}{m(m-1)} \right) + (g - 1). \quad (2.5)$$

If

$$m(f_m - f_{m-1}) \geq f_m, \quad (2.6)$$

---

[1]Note that the expression (2.2), as it appears in the proof of [2, Theorem 4.28] (lines 12 and 13 from the bottom of p. 558), has a misprint that we have corrected (r_q instead of d_q in the first term).

the expression (2.5) is non–negative, and so the largest fraction in (2.4) is $F(r-1)$. Moreover, the condition $d > F(r-1)$ would imply the inequalities (2.4) and then (2.2) as wanted. If we take

$$f_m = n\frac{f_{m-1}}{m-1} - (g-1),$$

and $f_1 = 2g - 1$, this recursion has a solution:

$$f_m = \frac{n^{m-1}}{(m-1)!}(2g-1) - \left(\frac{n^{m-1}}{(m-1)!}\sum_{k=1}^{m-1}\frac{k!}{n^k}\right)(g-1)$$

for any $m \geq 2$. This sequence is increasing in m and satisfies (2.6). Moreover $f_r \geq r(2g-2)$. Then, we can take $f(g,r,n) := f_r$ and the lemma holds.

Q.E.D.

As $\overline{M}_Q(d, n-r, n) = Q(n,r,d)$, in the statement of Theorem 1.1, we should take $b_1(g,r,n) := f(g, n-r, n) + 1$, with f as defined above.

3 Divisor class group of the Quot scheme

In this section, we compute the divisor class group of the Quot scheme $Q(n,r,d)$, for curves C of arbitrary genus.

The computation, as remarked in the introduction, uses [7, Proposition 1.8 & Theorem 3.3], and, through Bifet's result ([4]), the theorem of Bialynicki-Birula.

3.1 Reduction to the scheme of morphisms

Let $\Delta := Q(n,r,d) \setminus M(n,r,d)$ be the boundary of the compactification $Q(n,r,d)$ of $M(n,r,d)$. The information we need about Δ is contained in the proof of [1, Theorem 1.4] where it is proved that there is a stratification of the boundary

$$\Delta = \bigcup_{0 < m \leq d} B_m.$$

such that each stratum B_m, parameterizing quotients with torsion of degree m, is a fibre bundle on $M(n,r,d-m)$ with irreducible fibres of dimension $(n-r)m$.

Lemma 3.1. If $d > b_1(g,r,n)$ then, for $r > 1$ the codimension of Δ in $Q(n,r,d)$ is at least 2, and, for $r = 1$, Δ has an unique component of codimension one.

Proof. We use the bound $b_0(g,r,n) := f(g,n-r,n)$ computed in Lemma 2.1; note that $b_1(g,r,n) = b_0(g,r,n) + 1$. Let us estimate the dimension of B_m. We consider two cases.

Case 1. If $d-m > b_0$, the scheme $M(n, r, d-m)$ is irreducible by Lemma 2.1 and we have

$$
\begin{aligned}
\dim(B_m) &= n(d-m) - r(n-r)(g-1) + (n-r)m \\
&= nd - r(n-r)(g-1) - rm
\end{aligned}
$$

Then, if $r = 1$, the stratum B_1 has codimension 1, and the other strata are of higher codimension. For $r \geq 2$, all the strata are of codimension greater than one.

In addition, B_1 is irreducible because $M(n, 1, d-1)$ is irreducible and B_1 is a fibre bundle on it.

Case 2. If $d - m \leq b_0$, the scheme $M(n, r, d - m)$ can have dimension greater than expected (i.e., $(d - m)n - r(n - r)(g - 1)$) and we need a different argument. If b_0 is not an integer, we replace it by its integer part. Let us denote by $\tilde{b}_0$ the integer $b_0 + 1$.

Let $\tilde{m}$ be the integer, satisfying $0 < \tilde{m} \leq \tilde{b}_0$, defined as $\tilde{b}_0 - d + m$, and consider the proper subschemes $B'_{\tilde{m}}$ of the irreducible scheme $Q(n, r, \tilde{b}_0)$, that, therefore, have dimension at most

$$
\tilde{b}_0 n - r(n-r)(g-1) - 1 \,.
$$

They are fibre bundles on $M(n, r, \tilde{b}_0 - \tilde{m})$ with irreducible fibres of dimension $\tilde{m}(n - r)$. Then we get the bound

$$
\dim(M(n, r, \tilde{b}_0 - \tilde{m})) \leq \tilde{b}_0 n - r(n-r)(g-1) - 1 - \tilde{m}(n-r). \qquad (3.1)
$$

Returning to the computation of the dimension of $B_m \subset Q(n, r, d)$, we obtain, from (3.1), the bound

$$
\begin{aligned}
\dim(B_m) &\leq \tilde{b}_0 n - r(n-r)(g-1) - 1 - (\tilde{b}_0 - d + m)(n-r) + m(n-r) \\
&= nd - r(n-r)(g-1) - 1 - r(d - \tilde{b}_0)
\end{aligned}
$$

and, then, the codimension of B_m in $Q(n, r, d)$ is at least $(d - \tilde{b}_0)r + 1$. As $d - \tilde{b}_0 \geq 1$ we also get, in this case, $\operatorname{codim}(\Delta, Q(n, r, d)) \geq 2$. \hfill Q.E.D.

3.2 Filtration of Harder–Narasimhan

In our computation of the divisor class group of $Q(n, r, d)$ we are going to use that its singular locus is contained in a subvariety of codimension at least 2. This is a consequence of a result proved by Kirwan [10], that uses the notion of Harder–Narasimhan type. For the sake of completeness, we briefly recall the meaning of this concept.

Let E be a vector bundle, of rank r and degree d, on a curve C. The bundle E is semistable if for any vector subbundle E_1, with degree d_1 and rank r_1, we have

$$\frac{d_1}{r_1} \leq \frac{d}{r}.$$

Every vector bundle E has a canonical filtration, the Harder–Narasimhan filtration, $0 \subseteq F_1 \subseteq \cdots \subseteq F_s = E$ with semistable quotients $D_j = F_j/F_{j-1}$, such that

$$\frac{d_j}{r_j} > \frac{d_{j+1}}{r_{j+1}} \qquad \text{for } 1 \leq j \leq s - 1.$$

The degrees d_j and ranks r_j of the quotients D_j determine the type μ of the vector bundle E, defined as the vector $(d_1/r_1, \ldots, d_s/r_s)$ in which each ratio d_j/r_j appears r_j times. We will say that μ is a (r, d) type. Define the integer d_μ associated with the type μ as

$$d_\mu = \sum_{i>j} (r_i d_j - r_j d_i + r_i r_j (g - 1)).$$

The following result was proved by Kirwan using deformation theory [10, Lemma 3.9 & Corollary 6.6].

Proposition 3.2.

1. Given any positive integer k, if d is greater than or equal to

$$r\left(2g + \max\{k, \tfrac{1}{4}r^2 g\}\right),$$

   there exists a finite set $\mathcal{U}$ of (r, d) types such that if $\mu \notin \mathcal{U}$ then

$$d_\mu > k$$

   and if E is a vector bundle of type $\mu \in \mathcal{U}$, then

$$H^1(C, E) = 0.$$

2. The locus of points, in $M(n, r, d)$, corresponding to vector bundles of fixed type μ is contained in a subvariety of complex codimension at least

$$d_\mu - n^2 g.$$

Let M^0 be the open subset of $M(n, r, d)$ defined by

$$M^0 := \{q \in M(n, r, d) \mid H^1(C, N_q^\vee) = 0\} \subset M(n, r, d).$$

As the obstruction space at a point q can be identified with $H^1(N_q^\vee \otimes E_q)$, we see that M^0 is contained in the set of smooth points of $M(n, r, d)$.

Corollary 3.3. If $d \geq b_2(g, r, n) := (n - r)(2g + 2 + n^2 g)$, $M(n, r, d) \setminus M^0$ has codimension at least two in $M(n, r, d)$.

Proof. Proposition 3.2 applied to the vector bundle $N_q^\vee$, with $k \geq 2 + n^2 g$ and $d \geq (n - r)(2g + k)$, proves that $M(n, r, d) \setminus M^0$ has codimension ≥ 2 in $M(n, r, d)$.   Q.E.D.

3.3 Transversality

Consider now the following diagram of quotients of $\mathcal{O}_C^n$:

$$
\begin{array}{c}
\mathcal{O}_C^n \longrightarrow \mathcal{E} \longrightarrow 0 \\
\downarrow \\
\mathcal{O}_C^{n-r} \\
\downarrow \\
0
\end{array}
$$

where the vertical quotient $\mathcal{O}_C^{n-r}$ is a trivial vector bundle, $W \otimes \mathcal{O}_C$, for W a quotient vector space, of dimension $n - r$, of $V := H^0(\mathcal{O}_C^n)$; and the horizontal one, $\mathcal{E}$, is a sheaf of rank r and degree d. We can complete the diagram composing maps and taking kernels or cokernels:

$$
\tag{3.2}
\begin{array}{ccccc}
& 0 & & 0 & \\
& \downarrow & & \downarrow & \\
0 \longrightarrow & N' & \longrightarrow & \mathcal{O}_C^r & \\
& \downarrow & & \downarrow & \searrow \\
0 \longrightarrow & N & \longrightarrow & \mathcal{O}_C^n \xrightarrow{q} \mathcal{E} \longrightarrow 0 \\
& & \searrow_\Psi & \downarrow p & \downarrow \\
& & & \mathcal{O}_C^{n-r} \longrightarrow \mathcal{C} \longrightarrow 0 \\
& & & \downarrow & \downarrow \\
& & & 0 & 0
\end{array}
$$

Definition 3.4. We will say that the quotient q is generically transverse to the projection p if for general $x \in C$ the fibres $\mathbb{P}(\mathcal{E}_x)$ and $\mathbb{P}(\mathcal{O}_{C,x}^{n-r})$, of $\mathbb{P}(\mathcal{E})$ and $\mathbb{P}(\mathcal{O}_C^{n-r})$, do not intersect inside $\mathbb{P}(\mathcal{O}_{C,x}^n)$.

Here $\mathbb{P}(\cdot)$ denotes the projective bundle of hyperplanes in the fibres of $\mathcal{E}$, and, therefore, $\mathbb{P}(\mathcal{E})$ and $\mathbb{P}(\mathcal{O}_C^{n-r})$ inject into $\mathbb{P}(\mathcal{O}_C^n)$.

Lemma 3.5. The quotient q is generically transverse to the projection p if and only if the map Ψ, in diagram (3.2), is an injection of sheaves.

Proof. The definition of generical transversality is equivalent to $\mathcal{C}$ being a torsion sheaf. As the rank of N is $n - r$, $\mathcal{C}$ is torsion if and only if N' is also a torsion sheaf, but as N' is a subsheaf of a torsion free sheaf, $\mathcal{O}_C^r$, N' must be zero.

Q.E.D.

To prove the transversality lemma that we need (Lemma 3.7), we will use a result on the irreducibility of the fibres of certain equivariant morphisms. Specifically,

Lemma 3.6. Let X and Y be irreducible varieties acted on by a connected algebraic group G. Assume that the action of G on Y is transitive and that its stabilizers are irreducible. Then, the fibres of an equivariant map, f, from X to Y are irreducible of dimension $\dim(X) - \dim(Y)$.

Proof. To begin with, every component of the general fibre has dimension equal to $\dim(X) - \dim(Y)$ (cf. [13, Chapter I, Theorem 7]), but all the fibres are isomorphic, with isomorphisms given by multiplication by elements of the group G. It only remains to be proved that the fibres are irreducible.

Let $X_i \subset X$, $i = 1, 2, \ldots, k$, be the irreducible components of the fibre of f over a point $y_0 \in Y$, and consider the image M_i of the multiplication map $G \times X_i \overset{\mu_i}{\to} X$. The sets M_i are constructible, that is, finite disjoint unions of locally closed sets.

Let us consider the commutative diagram of G-equivariant maps

$$
\begin{array}{ccc}
G \times X_i & \overset{\mu_i}{\longrightarrow} & X \\
 & \searrow{\scriptstyle f_i} & \downarrow{\scriptstyle f} \\
 & & Y
\end{array}
$$

with G acting on $G \times X_i$ by multiplication on the first factor and trivially on the second.

The fibre, F_{y_0}, of f_i over the point $y_0 \in Y$ is isomorphic to $G_{y_0} \times X_i$, with G_{y_0} the stabilizer of y_0. Specifically, $G_{y_0} \times X_i$ injects into $G \times X_i$, and we need to check that the injection has image F_{y_0}: let (g_0, x_i) be a point in the fibre; we have $y_0 = f(g_0 \cdot x_i) = g_0 \cdot y_0$, and, therefore, $g_0 \in G_{y_0}$.

Now, we can identify a nonempty fibre of μ_i, F_{x_i}, over a point $x_i \in X_i \backslash \bigcup_{j \neq i} X_j$ and lying over y_0, with G_{y_0}. The point $(1, x_i) \in F_{y_0} = G_{y_0} \times X_i$ belongs to F_{x_i}. Let (h, x_i') be another point of F_{x_i}; this means that $h \cdot x_i' = x_i$ or $h^{-1} \cdot x_i = x_i'$. We define the map we need as

$$
\begin{array}{ccc}
G_{y_0} & \overset{\Phi}{\longrightarrow} & F_{y_0} = G_{y_0} \times X_i \\
h & \longmapsto & (h, h^{-1} \cdot x_i)
\end{array}
$$

and we still have to check that $h^{-1} \cdot x_i \in X_i$. In order to prove this, we observe that the orbit of x_i by G_{y_0} is irreducible, and then, containing a point $x_i \in X_i \setminus \bigcup_{j \neq i} X_j$, should be completely contained in X_i. The map Φ is an isomorphism between G_{y_0} and F_{x_i}, and the dimension of F_{x_i} equals $\dim(G) - \dim(Y)$.

Furthermore, the image M_i of μ_i must contain a nonempty open subset $U_i \subset X$; otherwise, the image would be contained in a proper closed subset, contradicting the existence of a fibre, F_{x_i}, of dimension equal to

$$\dim(G \times X_i) - \dim(X) = \dim(G) - \dim(Y).$$

Let $U \subset X$ be the intersection of U_1 and U_2, and X' the intersection of U with an irreducible component of a fibre of f. Then, there are, by construction, open sets $X'_i \subset X_i$, $i = 1, 2$, and elements $g_1, g_2 \in G$ such that $g_1 \cdot X'_1 = X' = g_2 \cdot X'_2$. This implies that $g_2^{-1} \cdot g_1(X'_1) = X'_2$, and then that $h := g_2^{-1} \cdot g_1 \in G_{y_0}$, but the image of X'_1 by multiplication by h has to be, as we have shown above, contained in X_1. This contradiction proves the lemma.   Q.E.D.

Let $M_{gt} \subset M(n, r, d)$ be the open subset parameterizing quotients generically transverse to a given projection $\mathcal{O}_C^n \xrightarrow{\mathrm{P}} \mathcal{O}_C^{n-r} \to 0$.

Lemma 3.7. For $d > b_1(n, r, g)$ we have

$$\mathrm{codim}(M(n, r, d) \setminus M_{gt}, M(n, r, d)) \geq 2.$$

Proof. Let us denote by $\mathrm{Gr}(r, n)$ the Grassmann variety of r-dimensional quotients of $\mathbb{K}^n$, and by $H \subset \mathrm{Gr}(r, n)$ its hyperplane section parameterizing quotients whose projectivization intersects $\mathbb{P}(W)$ nontrivially.

The linear group $\mathrm{GL}(n)$ acts on $M(n, r, d)$, transitively on $\mathrm{Gr}(r, n)$, and trivially on the curve C, and the natural map of evaluation

$$C \times M(n, r, d) \xrightarrow{\nu} \mathrm{Gr}(r, n),$$

is equivariant. Denote by $\Gamma \subset C \times M(n, r, d)$ the inverse image by ν of H. By Lemma 3.6, Γ is irreducible of codimension one.

The locus $M(n, r, d) \setminus M_{gt}$ is the set of points $m \in M(n, r, d)$ such that the dimension of $\pi_2^{-1}(m)$, with π_2 the projection of Γ over $M(n, r, d)$, is one. The map π_2, restricted to Γ, is generically finite of degree d because the fibre over $m \in M(n, r, d)$ consists of the pairs $(x, m) \in C \times M(n, r, d)$ such that $\nu(x, m) \in H$. Note that $\{x : \nu(x, m) \in H \text{ and } m \in M_{gt}\}$ is the support of the sheaf $\mathcal{C}$ in the diagram (3.2). Therefore, there is an exceptional locus $Z \subsetneq \Gamma$ of codimension at least one and such that the restriction of π_2 to Z has fibres of dimension one. Then, the image of Z by π_2, which coincides with the locus $M(n, r, d) \setminus M_{gt}$, has codimension at least two in $M(n, r, d)$.   Q.E.D.

3.4 The divisor class group of $Q(n, r, d)$

In order to finish the computation, we will need some results of [9] and [4]. Assume $d > \max\{b_1(g, r, n), b_2(g, r, n)\}$.

To begin with, let us fix a projection $\mathcal{O}_C^n \longrightarrow \mathcal{O}_C^{n-r} \to 0$. The transversality result, contained in the previous subsection, produces an open set $M_{gt} \subset M(n, r, d)$ parameterizing quotients generically transverse to the chosen projection. For each point $\bar{q} \in M_{gt}$ we get, by Lemma 3.5, an exact sequence

$$0 \to N_{\bar{q}} \longrightarrow \mathcal{O}_C^{n-r} \longrightarrow T_{\bar{q}} \to 0, \tag{3.3}$$

where $T_{\bar{q}}$ is a torsion sheaf of degree d, and hence is represented by a point of the scheme $\mathcal{Q} := Q(n-r, 0, d)$ parameterizing torsion quotients, of degree d, of $\mathcal{O}_C^{n-r}$. Note that $\mathcal{Q}$ is projective and smooth, of dimension $(n - r)d$.

Thus, by the universal property of $\mathcal{Q}$, we get a morphism $M_{gt} \overset{f}{\longrightarrow} \mathcal{Q}$. For each point $q \in \mathcal{Q}$ there is an exact sequence

$$0 \to N_q \overset{\varphi_q}{\longrightarrow} \mathcal{O}_C^{n-r} \longrightarrow T_q \to 0,$$

which coincides with (3.3) if $q = f(\bar{q})$, and we define $\mathcal{Q}^0 \subset \mathcal{Q}$ as the open set where $H^1(N_q^\vee) = 0$. This set is nonempty as soon as d is greater than $g(n - r)$.

Lemma 3.8. For $k \geq 2 + g(n - r)^2$ and $d \geq (n - r)(2g + k)$, we have that:

$$\mathrm{codim}(\mathcal{Q} \setminus \mathcal{Q}^0, \mathcal{Q}) \geq 2.$$

Proof. Just observe that $\mathcal{Q} \setminus \mathcal{Q}^0$ is the image by f of $M_{gt} \setminus M_{gt}^0$, that, by Corollary 3.3, is of codimension at least 2.   Q.E.D.

Now we see that $f^{-1}(\mathcal{Q}^0)$ can be identified with an open set, U, in a vector bundle, and the projection map of this vector bundle, restricted to U, coincides with f. Specifically, the vector bundle $\mathbb{V} \longrightarrow \mathcal{Q}^0$ has fibre over the point q isomorphic to $\mathrm{Hom}(N_q, \mathcal{O}_C^r)$, and a natural embedding, j, into $Q(n, r, d)$: to define j, at the point corresponding to an homomorphism $\psi_q \in \mathrm{Hom}(N_q, \mathcal{O}_C^r)$, take as image the point corresponding to the quotient, not necessarily locally free, of $N_q \overset{\psi_q \oplus \varphi_q}{\longrightarrow} \mathcal{O}_C^r \oplus \mathcal{O}_C^{n-r}$.

The situation can be summarized in the diagram given in Figure 1 where $\mathbb{V}$ and $M(n, r, d)$ are birationally equivalent through $\Upsilon := j|_{j^{-1}(M(n,r,d))}$.

Furthermore, using the notations M^0 (or: M_{gt}^0) to refer to the open set in $M(n, r, d)$ where $H^1(N_{\bar{q}}^\vee) = 0$ (or: $H^1(N_{\bar{q}}^\vee) = 0$ and the quotients are generically transverse to the chosen projection), we have

$$Q(n, r, d) \setminus \mathbb{V} \subset (Q(n, r, d) \setminus M(n, r, d)) \cup (M(n, r, d) \setminus M^0) \cup (M^0 \setminus M_{gt}^0)$$

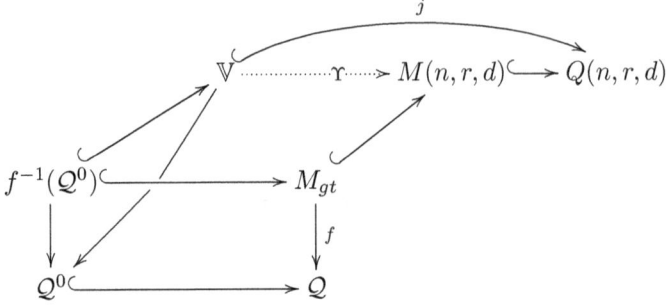

FIGURE 1.

and, for $d > \max\{b_1(g,r,n), b_2(g,r,n)\}$ and $r \neq 1$, the irreducible components of the sets in the right hand side are, by Lemma 3.1, Corollary 3.3 and Lemma 3.7, of codimension at least 2. This allows us to conclude that, if $r \neq 1$,

$$\mathrm{Cl}(Q(n,r,d)) = \mathrm{Pic}(\mathbb{V}). \tag{3.4}$$

If $r = 1$, $Q(n,r,d) \setminus M(n,r,d)$ has an irreducible component of codimension 1 having an open set B_1 parameterizing quotients with a torsion of degree 1. We claim that, assuming $n \geq 3$, B_1 has nonempty intersection with $\mathbb{V}$, and, thus $\mathbb{V}$ will contain an open set of B_1 and, again, we get $\mathrm{codim}(Q(n,r,d) \setminus \mathbb{V}, Q(n,r,d)) \geq 2$.

To prove the claim, let us begin with a general point $q \in M(n,1,d-1)^0$, parameterizing a quotient $\mathcal{O}_C^n \to L_q \to 0$ with kernel N_q, and corresponding to a generically injective map $C \xrightarrow{g_q} \mathbb{P}^{n-1}$.

Choose a general point $x \in C$ and a general line $\Lambda \subset \mathbb{P}^{n-1}$ containing the point $g_q(x)$. Now, restrict the Euler sequence of $\mathbb{P}^{n-1}$ to $C \cup \Lambda$ and take the direct image by the map from $C \cup \Lambda$ to C sending Λ to x. We get in this way an exact sequence on C

$$0 \to N' \to \mathcal{O}_C^n \to L_q \oplus T \to 0, \tag{3.5}$$

with T a torsion sheaf of degree 1 supported on x, and such that composing the quotient with the projection to L_q we get the quotient parameterized by q. The kernel N' injects into N_q, with quotient T, and, then, we get, taking cohomology, $H^1((N')^\vee) = 0$.

Finally, we change, if necessary, the projection $\mathcal{O}_C^n \xrightarrow{p} \mathcal{O}_C^{n-1} \to 0$ in order to get that the quotient $q' \in B_1$ given by the sequence (3.5) is generically transverse to the new projection. The change of p will change the open set M_{gt}, but does not change the vector bundle $\mathbb{V}$ nor the map j. Then, q' is in the image by j of $\mathbb{V}$, and the claim has been proved.

Moreover, the proof of Theorem 1.1 has been reduced to the computation of the Picard group of the vector bundle $\mathbb{V} \to \mathcal{Q}^0$.

Recall that $\mathcal{Q}$ is $Q(n-r,0,d)$.

Lemma 3.9. We have, for $d > 2g+1$,

$$\text{Pic}(\mathbb{V}) = \mathbb{Z} \oplus \mathbb{Z} \oplus \text{Pic}(\text{Jac}^d(C)) \quad \text{if} \quad 1 \leq r < n-1 \qquad (3.6)$$

Proof. We know, by Lemma 3.8 and applying [7, Thm. 3.3], that

$$\text{Pic}(\mathbb{V}) = \text{Pic}(\mathcal{Q}^0) = \text{Pic}(\mathcal{Q}).$$

In order to compute $\text{Pic}(\mathcal{Q})$, we use [4], where a decomposition of $\mathcal{Q}$ is obtained (using [3]) with each stratum isomorphic to a vector bundle, of known rank, on a product of symmetric powers of C.

Claim [4, Corollary, p. 611]. Let C be a projective smooth curve. There is a smooth decomposition $(\mathcal{S}_{\underline{d}})$ of $\mathcal{Q}$, indexed by the partitions $\underline{d} = (d_1, \ldots, d_{n-r})$ of d. Each stratum $\mathcal{S}_{\underline{d}}$ is isomorphic to a vector bundle on $C^{(d_1)} \times \cdots \times C^{(d_{n-r})}$ and it has rank $r_{\underline{d}} = \sum_{1 \leq i \leq n-r}(i-1)d_i$.

As a consequence, we have in $\mathcal{Q}$ a dense open $\mathcal{U} := \mathcal{S}_{\underline{d}}$, for $\underline{d} = (0, \ldots, 0, d)$, which is a vector bundle on $C^{(d)}$. Thus (for $d > 2g+1$)

$$\text{Pic}(\mathcal{U}) = \text{Pic}(C^{(d)}) = \mathbb{Z} \oplus \text{Pic}(\text{Jac}^d(C)).$$

Moreover, if $r \neq n-1$, there is a unique stratum in the decomposition of $\mathcal{Q}$ having codimension 1, namely

$$Z := \mathcal{S}_{\underline{d}}, \quad \text{for } \underline{d} = (0, \ldots, 1, d-1).$$

Then, the Picard group of $\mathcal{Q}$ is in an exact sequence (cf. [7, Proposition 1.8])

$$0 \to \mathbb{Z}[\overline{Z}] \to \text{Pic}(\mathcal{Q}) \to \text{Pic}(\mathcal{U}) \to 0, \qquad (3.7)$$

that is exact on the left because, by projectivity of the variety $\mathcal{Q}$, there are no effective torsion classes in its Picard group.

Furthermore, we claim that the sequence splits. To prove the claim just observe that the second and third groups in the sequence contain isomorphic copies of $\text{Pic}(\text{Jac}^d(C))$:

$$0 \twoheadrightarrow \mathbb{Z}[\overline{Z}] \rightarrowtail \text{Pic}(\mathcal{Q}) \longrightarrow \text{Pic}(\mathcal{U}) \to 0$$

$$\pi^* \searrow \qquad \nearrow \pi^*$$

$$\text{Pic}(\text{Jac}^d(C))$$

with π the natural map from $\mathcal{Q}$ to the Jacobian, sending each quotient to the top exterior power of its kernel dualized. Taking the quotient, in both groups, by $\mathrm{Pic}(\mathrm{Jac}^d(C))$, and using $\mathrm{Pic}(\mathcal{U}) = \mathbb{Z} \oplus \mathrm{Pic}(\mathrm{Jac}^d(C))$, we get

$$
\begin{array}{ccc}
0 & & 0 \\
\downarrow & & \downarrow \\
\mathrm{Pic}(\mathrm{Jac}^d(C)) & \!\!=\!\!=\!\!=\!\! & \mathrm{Pic}(\mathrm{Jac}^d(C)) \\
\downarrow & & \downarrow \\
0 \to \mathbb{Z}[\overline{Z}] \to \mathrm{Pic}(\mathcal{Q}) \to \mathrm{Pic}(\mathcal{U}) \to 0 \\
\| \qquad \downarrow \qquad \downarrow \\
0 \to \mathbb{Z} \to \mathrm{Pic}(\mathcal{Q})/\mathrm{Pic}(\mathrm{Jac}^d(C)) \to \mathbb{Z} \to 0 \\
\downarrow \qquad \downarrow \\
0 \qquad 0
\end{array}
$$

with the sequence at the bottom split. Now, the extension class of the vertical sequence on the right is zero, and, therefore, the class of the vertical middle sequence is also zero.   Q.E.D.

This completes the proof of Theorem 1.1. The proof of the theorem also shows that

$$
\mathrm{Cl}(M(n,r,d)) = \begin{cases} \mathbb{Z} \oplus \mathrm{Pic}(\mathrm{Jac}^d(C)) & \text{if} \quad r = 1,\, n-1 \\ \mathbb{Z} \oplus \mathbb{Z} \oplus \mathrm{Pic}(\mathrm{Jac}^d(C)) & \text{if} \quad 1 < r < n-1 \end{cases} \tag{3.8}
$$

if $d > \max\{b_1(g,r,n),\, b_2(g,r,n)\}$.

References

[1] A. Bertram. Towards a Schubert calculus for maps from a Riemann surface to a Grassmannian. *International Journal of Mathematics*, 5(6):811–825, 1994.

[2] A. Bertram, G. Daskalopoulos, and R. Wentworth. Gromov invariants for holomorphic maps from Riemann surfaces to Grassmannians. *Journal of the American Mathematical Society*, 9(2):529–571, 1996.

[3] A. Białynicki-Birula. Some theorems on actions of algebraic groups. *Annals of Mathematics*, 98:480–497, 1973.

[4] E. Bifet. Sur les points fixes du schéma $\mathrm{Quot}_{\mathcal{O}^r_{X/X/k}}$ sous l'action du tore $\mathbf{G}^r_{m,k}$. *Comptes Rendus de l'Académie des Sciences. Série I. Mathématique*, 309(9):609–612, 1989.

[5] J.-M. Drezet and M. S. Narasimhan. Groupe de Picard des variétés de modules de fibrés semi-stables sur les courbes algébriques. *Inventiones Mathematicae*, 97(1):53–94, 1989.

[6] J. Fogarty. Algebraic families on an algebraic surface. II. The Picard scheme of the punctual Hilbert scheme. *American Journal of Mathematics*, 95:660–687, 1973.

[7] W. Fulton. *Intersection theory*, volume 2 of *Ergebnisse der Mathematik und ihrer Grenzgebiete*. Springer-Verlag, Berlin, 2nd edition, 1998.

[8] A. Grothendieck. Techniques de construction et théorèmes d'existence en géométrie algébrique. IV. Les schémas de Hilbert. In *Séminaire Bourbaki. Volume 6: Année 1960/61*, pages 249–276. Societé Mathématiques de France, Paris, 1995. Exposé 221.

[9] R. Hernández. On Harder-Narasimhan stratification over Quot schemes. *Journal für die Reine und Angewandte Mathematik*, 371:115–124, 1986.

[10] F. Kirwan. On spaces of maps from Riemann surfaces to Grassmannians and applications to the cohomology of moduli of vector bundles. *Arkiv för Matematik*, 24(2):221–275, 1986.

[11] A. Kouvidakis. Picard groups of Hilbert schemes of curves. *Journal of Algebraic Geometry*, 3(4):671–684, 1994.

[12] M. Popa and M. Roth. Stable maps and Quot schemes. *Inventiones Mathematicae*, 152(3):625–663, 2003.

[13] I. R. Shafarevich. *Basic algebraic geometry. 1*. Springer-Verlag, Berlin, 2nd edition, 1994. Varieties in projective space, Translated from the 1988 Russian edition and with notes by Miles Reid.

[14] S. A. Strømme. On parametrized rational curves in Grassmann varieties. In F. Ghione, C. Peskine, and E. Sernesi, editors, *Space curves. Proceedings of the conference on curves in projective space held in Rocca di Papa, June 3–8, 1985*, volume 1266 of *Lecture Notes in Mathematics*, pages 251–272, Berlin, 1987. Springer-Verlag.

[15] E. Viehweg. *Quasi-projective moduli for polarized manifolds*, volume 30 of *Ergebnisse der Mathematik und ihrer Grenzgebiete*. Springer-Verlag, Berlin, 1995.

Spectrally compact operators

Shirin Hejazian, Mohadeseh Rostamani*

Department of Pure Mathematics, Ferdowsi University of Mashhad, P.O. Box 1159,
Mashhad 91775, Iran

E-mail: hejazian@um.ac.ir, mohadeseh.rostamani@gmail.com

Abstract

We define the concept of a spectrally compact operator, and study
the basic properties of these operators. We show that the class of
spectrally compact operators is strictly contained in the class of compact operators and in the class of spectrally bounded operators. It is
also proved that the set of spectrally compact operators on a spectrally normed space E is a right ideal of $\mathrm{SB}(E)$ and in certain cases
it is a two sided ideal. We will also study the spectral adjoint of a
spectrally compact operator.

2000 Mathematics Subject Classification. **47B48**. 46B99, 47L10.
Keywords. spectrally normed space, spectrally bounded operator, spectrally compact operator.

1   Introduction

Let E be a normed space endowed with a spectral structure in the sense
that there exists a linear topological isomorphism τ from E into a unital
Banach algebra A. We consider E as a normed subspace of A, and we
write $E \subseteq A$. Such a normed space E is said to be a *spectrally normed*
space. It should be emphasized that the spectral structure on E depends
on the embedding, up to topological isomorphisms. For $x \in E$, $\mathrm{sp}(x)$ and
$\mathrm{r}(x)$ denote the spectrum and the spectral radius of x with respect to the
Banach algebra A, respectively. A spectrally normed space is said to be
commutative (semisimple) whenever A is commutative (semisimple).

Every normed space E carries at least one spectral structure via the iso-
metric embedding $j_E : E \to \mathrm{C}(E_1{}^*)$, the complex-valued continuous func-
tions on the dual closed unit ball $E_1{}^*$, endowed with the weak*-topology.
This is a commutative semisimple structure and $\|x\| = \mathrm{r}(x)$ for $x \in E$.

Let E, F be spectrally normed spaces. A linear mapping $T : E \to F$ is
called *spectrally bounded*, if there exists $M \geq 0$, such that $\mathrm{r}(Tx) \leq M\mathrm{r}(x)$,
for all $x \in E$. In general, a spectrally bounded operator need not be bounded
and conversely, a bounded operator between spectrally normed spaces may
not be spectrally bounded, see [3, Examples 2.7, 2.8].

---

*The authors would like to express their deepest thanks to the referees for valuable
comments and suggesting the shorter proof of Theorem 2.10.

Tbilisi Mathematical Journal 3 (2010), pp. 17–25.
Tbilisi Centre for Mathematical Sciences & College Publications.

Received by the editors: 27 March 2010; 13 October 2010.
Accepted for publication: 22 November 2010.

Mathieu and Schick initiated a systematic study of spectrally bounded operators between spectrally normed spaces in [3]. Spectrally bounded operators on von Neumann algebras and simple C^*-algebras were studied in [4, 2].

For each pair of spectrally normed spaces E and F, we denote by $B(E,F), K(E,F)$ and $SB(E,F)$, the space of all bounded operators, all compact operators and all spectrally bounded operators from E to F, respectively. The closed unit ball of E is denoted by E_1. For $T \in SB(E,F)$ the value

$$\|T\|_\sigma = \inf\{M \geq 0 : r(Tx) \leq Mr(x), x \in E\},$$

is called the *spectral operator norm* of T. We recall the following results from [3].

Proposition 1.1. [3, Proposition 2.4] Let E, F and G be spectrally normed spaces and $S, T \in SB(E,F)$ and $R \in SB(F,G)$ then

1. $\|T\|_\sigma = \sup\{r(Tx) : x \in E, r(x) \leq 1\} = \sup\{r(Tx) : x \in E, r(x) = 1\}$;

2. $\|\lambda T\|_\sigma = |\lambda|\|T\|_\sigma$ for all $\lambda \in \mathbb{C}$;

3. $\|RT\|_\sigma \leq \|R\|_\sigma\|T\|_\sigma$;

4. $\|S + T\|_\sigma \leq \|S\|_\sigma + \|T\|_\sigma$, if F is commutative.

Proposition 1.2. [3, Proposition 2.5] Suppose that F is a commutative semisimple spectrally normed space. For every spectrally normed space E, $(SB(E,F), \|.\|_\sigma)$ is a normed space. If $E = F$ then $SB(E) = SB(E,E)$ is a unital normed algebra.

In Section 2, we define spectrally compact operators and study some of their basic properties. We show that the class of spectrally compact operators is strictly contained in the class of compact operators and in the class of spectrally bounded operators. We will also show that the set of spectrally compact operators on a spectrally normed space E is a right ideal of the algebra $SB(E)$ which, in certain cases, is a two sided ideal. Section 3 is devoted to the study of the spectral adjoint of a spectrally compact operator. We will show that the spectral adjoint of every spectrally compact operator is spectrally compact, but it remains open if this is in fact an equivalence.

2   Basic properties of spectrally compact operators

From now on, throughout the paper, E and F are assumed to be complex spectrally normed spaces.

Definition 2.1. A linear mapping $T : E \to F$ is said to be *spectrally compact* if $\overline{T(U_E)}$ is compact in F, where $U_E = \{x \in E : r(x) \leq 1\}$ and

E, F are spectrally normed spaces. The set of spectrally compact operators from E to F is denoted by $\mathrm{KSB}(E, F)$.

Proposition 2.2. If $T : E \to F$ is a spectrally compact operator, then T is spectrally bounded.

Proof. Let $U_E = \{x \in E : \mathrm{r}(x) \leq 1\}$. Since $\overline{T(U_E)}$ is compact, there is $M \geq 0$ such that

$$\|T(x)\| \leq M \text{ for every } x \in U_E. \tag{2.1}$$

Now suppose that $x \in E$. If $\mathrm{r}(x) > 0$ then by (2.1), $\|T(\frac{x}{\mathrm{r}(x)})\| \leq M$ and so $\mathrm{r}(Tx) \leq M\mathrm{r}(x)$. If $\mathrm{r}(x) = 0$, for given $\varepsilon > 0$ we have $\mathrm{r}(\frac{x}{(\varepsilon/M)}) = 0$. Thus by (2.1)

$$\left\| T\left(\frac{x}{(\varepsilon/M)} \right) \right\| \leq M$$

and $\|T(x)\| \leq M \cdot \dfrac{\varepsilon}{M} = \varepsilon$. Therefore $\mathrm{r}(Tx) \leq \|Tx\| = 0$, since $\varepsilon > 0$ is arbitrary. It follows that $\mathrm{r}(Tx) \leq M\mathrm{r}(x)$ for all $x \in E$, and hence $T \in \mathrm{SB}(E, F)$.   Q.E.D.

Proposition 2.3. Each spectrally compact operator $T : E \to F$ is a compact operator.

Proof. Let E_1 be the closed unit ball of E. Since for each $x \in E_1$, $\mathrm{r}(x) \leq 1$ we have $E_1 \subseteq U_E$, and $\overline{T(E_1)} \subseteq \overline{T(U_E)}$. Therefore $\overline{T(E_1)}$ is compact.   Q.E.D.

The following examples show that in Propositions 2.2 and 2.3 the reverse implications do not hold.

Example 2.4. (i) Let A be an infinite dimensional commutative unital Banach algebra and $M_2(\mathbb{C})$ the C^*-algebra of all complex 2×2 matrices. Suppose that φ is a character and f is an unbounded linear functional on A. Then the linear mapping $T : A \to M_2(\mathbb{C})$ defined by

$$T(a) = \begin{pmatrix} \varphi(a) & f(a) \\ 0 & \varphi(a) \end{pmatrix} \text{ for every } a \in A,$$

is an unbounded operator and for each $a \in A$, $\mathrm{r}(T(a)) = |\varphi(a)| \leq \mathrm{r}(a)$. Thus T is spectrally bounded. Clearly this mapping is not spectrally compact, otherwise, by Proposition 2.3, it should be compact and hence bounded.

(ii) Let $A = M_2(\mathbb{C})$, and $f : A \to \mathbb{C}$ a linear functional defined by $f(a_{ij}) = a_{12}$ for all $(a_{ij}) \in A$. Clearly, f is compact. We show that it is not spectrally compact. Let

$$a = \begin{pmatrix} 0 & 1 \\ 0 & 0 \end{pmatrix},$$

then $r(a) = 0$ and $r(f(a)) = 1$. It follows that f is not spectrally bounded and hence it is not spectrally compact.

If E and F are spectrally normed spaces and $T : E \to F$ is a linear mapping then it is easy to see that for $R \geq 0$,

(i) $\overline{\{T(x) : r(x) \leq R\}}$ is compact if and only if $\overline{\{T(x) : r(x) \leq 1\}}$ is compact.

(ii) $\overline{\{T(x) : r(x) < R\}}$ is compact if and only if $\overline{\{T(x) : r(x) < 1\}}$ is compact.

A subset B of a spectrally normed space E is called *spectrally bounded* if there is $M \geq 0$ such that $r(x) \leq M$, for all $x \in B$.

Theorem 2.5. Suppose that $T : E \to F$ is a linear mapping between spectrally normed spaces E and F. Then the following conditions are equivalent:

(i) T is a spectrally compact operator,

(ii) for every spectrally bounded subset B of E, $\overline{T(B)}$ is compact, and

(iii) for every spectrally bounded sequence (x_n) in E, $(T(x_n))$ has a convergent subsequence in F.

Proof. "(i)$\Leftrightarrow$(ii)" is obvious.

"(ii)$\Rightarrow$(iii)". Suppose that (x_n) is a spectrally bounded sequence in E. Then there is $M \geq 0$ such that $r(x_n) \leq M$ for all $n \in \mathbb{N}$. Let $B = \{x \in E : r(x) \leq M\}$. By (ii), $\overline{T(B)}$ is compact and $(T(x_n))$ is a sequence in the compact set $\overline{T(B)}$, so it has a convergent subsequence.

"(iii)$\Rightarrow$(ii)". Suppose that B is a spectrally bounded subset of E. To show that $\overline{T(B)}$ is compact, we prove that every sequence in this set has a convergent subsequence. Let (y_n) be a sequence in $\overline{T(B)}$, then there exists a sequence (x_n) in B such that

$$\|y_n - Tx_n\| < \frac{1}{n} \text{ for every } n \in \mathbb{N}.$$

Since (x_n) is a spectrally bounded sequence, by the hypothesis there is a subsequence (Tx_{n_j}) of (Tx_n), such that $Tx_{n_j} \to y$ for some $y \in F$, as $j \to \infty$. Let $\varepsilon > 0$ be given. There are $N_1, N_2 \in \mathbb{N}$ such that $\frac{1}{n_j} < \frac{\varepsilon}{2}$ for all $j \geq N_1$ and

$$\|Tx_{n_j} - y\| < \frac{\varepsilon}{2} \text{ for every } j \geq N_2.$$

So for all $j \geq \max(N_1, N_2)$ we have

$$\|y_{n_j} - y\| \leq \|y_{n_j} - Tx_{nj}\| + \|Tx_{nj} - y\| < \frac{\varepsilon}{2} + \frac{\varepsilon}{2} = \varepsilon.$$

Thus $y_{n_j} \to y$.   Q.E.D.

Corollary 2.6. If $S, T : E \to F$ are spectrally compact operators, and if $\lambda \in \mathbb{C}$ then $S + \lambda T$ is a spectrally compact operator.

As a consequence of Proposition 2.2 and Corollary 2.6, $\mathrm{KSB}(E, F)$ is a linear subspace of $\mathrm{SB}(E, F)$.

Theorem 2.7. Let X be a dense linear subspace of E and $T : X \to F$ a spectrally compact operator. Then T has a unique spectrally compact extension $\tilde{T} : E \to F$.

Proof. Since T is bounded we can extend T to a bounded operator $\tilde{T} : E \to F$. Suppose (x_n) is a spectrally bounded sequence in E and let $M \geq 0$ be such that $\mathrm{r}(x_n) \leq M$ for all $n \in \mathbb{N}$. Suppose $V = \{\lambda \in \mathbb{C} : |\lambda| < M\}$, then $\mathrm{sp}(x_n) \subseteq V$ for all $n \in \mathbb{N}$. Since $x_1 \in \overline{X}$, there is a sequence (x_{1n}) in X such that $x_{1n} \to x_1$. By [1, Theorem 3.4.2], there exists $0 < \delta_1 < 1$ such that $\|x_1 - y\| < \delta_1$ implies that $\mathrm{sp}(y) \subseteq V$, that is $\mathrm{r}(y) \leq M$. Take $\varepsilon_1 < \delta_1$, then there exists $n_1 \in \mathbb{N}$ such that

$$\|x_{1n} - x_1\| < \varepsilon_1 < 1 \text{ for every } n \geq n_1.$$

Thus $\mathrm{sp}(x_{1n}) \subseteq V$ for all $n \geq n_1$. Similarly for x_2 there exists a sequence (x_{2n}) in X such that $x_{2n} \to x_2$. Again by [1, Theorem 3.4.2], there exists $0 < \delta_2 < \frac{1}{2}$ such that $\|x_2 - y\| < \delta_2$ implies that $\mathrm{sp}(y) \subseteq V$. Take $0 < \varepsilon_2 < \delta_2$, there exists $n_2 \in \mathbb{N}$ such that

$$\|x_{2n} - x_2\| < \varepsilon_2 < \frac{1}{2} \text{ for every } n \geq n_2.$$

Therefore $\mathrm{sp}(x_{2n}) \subseteq V$ for all $n \geq n_2$. An inductive argument gives us a sequence $(y_k) \subseteq X$ $(y_k = x_{kn_k}(k \in \mathbb{N}))$ such that $x_k - y_k \to 0$ and $\mathrm{sp}(y_k) \subseteq V$. Since T is spectrally compact, there is a subsequence (y_{k_j}) such that $T y_{k_j} \to y_0$ for some $y_0 \in F$, as $j \to \infty$. Since $\tilde{T}$ is continuous

$$\tilde{T} x_{k_j} - \tilde{T} y_{k_j} = \tilde{T}(x_{k_j} - y_{k_j}) \to 0.$$

So $\tilde{T} x_{k_j} \to y_0$ and hence $\tilde{T}$ is spectrally compact. Q.E.D.

Proposition 2.8. If $T \in \mathrm{SB}(E, F)$ and $S \in \mathrm{KSB}(F, G)$ for some spectrally normed spaces E, F and G, then $ST : E \to G$ is spectrally compact.

Proof. Suppose that (x_n) is a spectrally bounded sequence in E, that is there exists $M \geq 0$ such that $\mathrm{r}(x_n) \leq M$ for all $n \in \mathbb{N}$. Since $T \in \mathrm{SB}(E, F)$, by Proposition 1.1

$$\mathrm{r}(T x_n) \leq \|T\|_\sigma \mathrm{r}(x_n) \leq M \|T\|_\sigma \text{ for every } n \in \mathbb{N}.$$

So $(T x_n)$ is a spectrally bounded sequence in F and hence it has a subsequence $(T x_{n_j})$ such that $(ST x_{n_j})$ converges, because S is spectrally compact. Therefore $ST \in \mathrm{KSB}(E, G)$. Q.E.D.

Corollary 2.9. $\mathrm{KSB}(E)$ is a right ideal of $\mathrm{SB}(E)$, and if $E \subseteq \mathrm{C}(Y)$ for some compact Hausdorff space Y, then $\mathrm{KSB}(E)$ is a two sided ideal of $\mathrm{SB}(E)$.

Proof. By Proposition 2.8, $\mathrm{KSB}(E)$ is a right ideal of $\mathrm{SB}(E)$. Suppose that $E \subseteq \mathrm{C}(Y)$, $T \in \mathrm{SB}(E)$ and $S \in \mathrm{KSB}(E)$. Let (x_n) be a spectrally bounded sequence in E. By Theorem 2.5, it has a subsequence (x_{n_j}) such that $Sx_{n_j} \to y$ for some $y \in E$, as $j \to \infty$. Since $E \subseteq \mathrm{C}(Y)$, by [3, Proposition 2.9], T is bounded and hence $T(Sx_{n_j}) \to Ty$. Thus TS is spectrally compact by Theorem 2.5. Q.E.D.

Theorem 2.10. Let E and F be spectrally normed spaces with $F \subseteq \mathrm{C}(Y)$, where Y is a compact Hausdorff space. The following statements hold:

  (i) The inclusion mapping $\iota : \mathrm{SB}(E, F) \to \mathrm{B}(E, F)$ is contractive,

 (ii) if there exists $M \geq 0$ satisfying $\|x\| \leq M$, for every x in $U_E = \{x \in E : \mathrm{r}(x) \leq 1\}$, the inequality $\|\iota(T)\| \leq \|T\|_\sigma \leq M\|\iota(T)\|$ holds for every T in $\mathrm{SB}(E, F)$,

(iii) $\mathrm{SB}(E, F)$ is a Banach space whenever F is a closed subspace of $\mathrm{C}(Y)$,

(iv) under the hypothesis in (ii), ι has closed range whenever F is a closed subspace of $\mathrm{C}(Y)$,

 (v) $\iota(\mathrm{KSB}(E, F) \subseteq \mathrm{K}(E, F)$ and under the assumptions in (ii), $\mathrm{KSB}(E, F)$ is a $\|.\|_\sigma$-closed subspace of $\mathrm{K}(E, F)$.

Proof. (i) Since $F \subseteq \mathrm{C}(Y)$, by [3, Proposition 2.9], we have

$$\mathrm{SB}(E, F) \subseteq \mathrm{B}(E, F),$$

and $\|T\| \leq \|T\|_\sigma$ for every $T \in \mathrm{SB}(E, F))$. Thus the inclusion mapping $\iota : \mathrm{SB}(E, F) \to \mathrm{B}(E, F)$ is contractive.

(ii) By hypothesis, $\|x\| \leq M$ for every $x \in U_E$. Therefore, given $x \in U_E$ and T in $\mathrm{SB}(E, F)$, we have

$$\mathrm{r}(T(x)) \leq \|T(x)\| \leq \|\iota(T)\|M.$$

Since x was arbitrarily chosen in U_E, we deduce, via Proposition 1.2 or [3, Proposition 2.4], that $\|T\|_\sigma \leq M\|\iota(T)\|$.

(iii) Let (T_n) be a $\|.\|_\sigma$-Cauchy sequence in $\mathrm{SB}(E, F)$. In this case, there exists $M \geq 0$ satisfying $\|T_n\|_\sigma \leq M$, for every $n \in \mathbb{N}$. Since $F \subseteq \mathrm{C}(Y)$, by [3, Proposition 2.9], we have $\mathrm{SB}(E, F) \subseteq \mathrm{B}(E, F)$ and $\|.\| \leq \|.\|_\sigma$ on $\mathrm{SB}(E, F)$. We deduce that every $\|.\|_\sigma$-Cauchy sequence in $\mathrm{SB}(E, F)$ is a $\|.\|$-Cauchy sequence in $\mathrm{B}(E, F)$. It follows that there exists $T \in \mathrm{B}(E, F)$ such that $\|T_n - T\| \to 0$ (F is a Banach space).

Let us fix an arbitrary $x \in E$. Since F is commutative the spectral radius is subadditive. Thus, the inequality

$$\begin{aligned} \mathrm{r}(T(x)) &\leq \mathrm{r}((T - T_n)(x)) + \mathrm{r}(T_n(x)) \\ &\leq \|T_n - T\|\|x\| + \|T_n\|_\sigma \mathrm{r}(x) \\ &\leq \|T_n - T\|\|x\| + M\mathrm{r}(x) \end{aligned}$$

holds for every $n \in \mathbb{N}$. Taking limit in $n \to \infty$ we have $\mathrm{r}(T(x)) \leq M\mathrm{r}(x)$. The required statement follows because x is arbitrary.

(iv) Follows from (i), (ii), and (iii).

(v) Proposition 2.5 above implies that $\iota(\mathrm{KSB}(E, F)) \subseteq \mathrm{K}(E, F)$. Now suppose that $\|T_n - T\|_\sigma \to 0$ where $(T_n) \subseteq \mathrm{KSB}(E, F)$ and $T \in \mathrm{SB}(E, F)$. It follows from (i) that $\|\iota(T_n) - \iota(T)\| \to 0$ in $\mathrm{B}(E, F)$. The sequence $(\iota(T_n))$ lies in $\mathrm{K}(E, F)$. Therefore, $\iota(T)$ is a compact operator and, since U_E is bounded, there exists $M \geq 0$ such that $\overline{T(U_E)} = \iota(T)(U_E) \subseteq \iota(T)(M(E_1))$ is compact. This shows that $T \in \mathrm{KSB}(E, F)$. Q.E.D.

Corollary 2.11. If $E = F \subseteq \mathrm{C}(Y)$ for a compact Hausdorff space Y, then $\mathrm{KSB}(E)$ is a closed two sided ideal of $\mathrm{SB}(E)$.

3 Spectral adjoint

We recall the following definition from [3]:

Definition 3.1. For a spectrally normed space E, $(\mathrm{SB}(E, \mathbb{C}), \|.\|_\sigma)$ is called the *spectral dual* of E and is denoted by E^σ.

Remark 3.2. By Proposition 1.2, for every spectrally normed space E, the space E^σ is normed. In fact it is a Banach space, see [3, Proposition 3.2]. If E^* denote the dual of E then by Theorem 2.10 (i) or [3, Proposition 2.9], we have a contractive embedding from E^σ into E^*. In other words, $E^\sigma \subseteq E^*$ and $\|f\| \leq \|f\|_\sigma$ for every $f \in E^\sigma$.

We consider E^σ as a spectrally normed space via the spectral structure inherited from the embedding into $\mathrm{C}((E^\sigma)_1^*)$. Moreover, in this spectral structure $\|f\|_\sigma = \mathrm{r}(f)$ for all $f \in E^\sigma$.

If $T : E \to F$ is a spectrally bounded operator, the linear operator $T^\sigma : F^\sigma \to E^\sigma$ defined by $T^\sigma g = g \circ T$ (for $g \in F^\sigma$), is said to be the *spectral adjoint* of T. The following proposition is [3, Corollary 3.7].

Proposition 3.3. The spectral adjoint T^σ of a spectrally bounded operator T is a spectrally bounded operator with $\|T^\sigma\|_\sigma \leq \|T\|_\sigma$.

We show that a similar result holds for spectrally compact operators.

Theorem 3.4. Suppose that $T : E \to F$ is a spectrally compact operator then $T^{\sigma} : F^{\sigma} \to E^{\sigma}$ is spectrally compact.

Proof. We consider a spectral structure on F^{σ} as in Remark 3.2. Let $B \subseteq F^{\sigma}$ be a spectrally bounded set, so there exists $C > 0$ such that

$$\|g\|_{\sigma} = \mathrm{r}(g) \leq C \text{ for every } g \in B.$$

We shall show that $T^{\sigma}(B)$ is totally bounded. Let $\varepsilon > 0$ be given. Since T is spectrally compact, $T(U_E) = \{Tx : \mathrm{r}(x) \leq 1\}$ is relatively compact and hence is totally bounded in F. So there are $x_1, \ldots, x_n \in U_E$ such that for every $x \in U_E$ there exists $j \in \{1, \ldots, n\}$ for which

$$\|Tx - Tx_j\| < \frac{\varepsilon}{3C}.$$

We define a linear operator $S : F^{\sigma} \to \mathbb{C}^n$ by

$$S(g) = (g(Tx_1), ..., g(Tx_n)) \text{ for every } g \in F^{\sigma}.$$

Since each $g \in B$ is a bounded linear functional and T is compact, S is a compact operator and hence $\overline{S(B)}$ is a compact set. Therefore $S(B)$ is totally bounded, that is there exist $g_1, \ldots, g_m$ in B such that for each $g \in B$ there exists $k \in \{1, \ldots, m\}$ with

$$\|Sg - Sg_k\|_e < \frac{\varepsilon}{3},$$

where $\|.\|_e$ denotes the Euclidean norm on $\mathbb{C}^n$. Thus for every $g \in B$ there is $k \in \{1, \ldots, m\}$ such that

$$|g(Tx_j) - g_k(Tx_j)|^2 \leq \sum_{\ell=1}^{n} |g(Tx_\ell) - g_k(Tx_\ell)|^2 = \|S(g - g_k)\|_e^2 < \frac{\varepsilon^2}{3^2},$$

for every $j \in \{1, \ldots, n\}$. Fix arbitrary $x \in U_E$ and $g \in B$. There is $j \in \{1, \ldots, n\}$ such that

$$\|Tx - Tx_j\| < \frac{\varepsilon}{3C}.$$

There is $k \in \{1, \ldots, m\}$ such that

$$\|Sg - Sg_k\|_e < \frac{\varepsilon}{3}.$$

For every $g \in B$ we have $\|g\| \leq \|g\|_{\sigma} \leq C$, thus

$$\begin{aligned}
|g(Tx) - g_k(Tx)| &\leq |g(Tx) - g(Tx_j)| + |g(Tx_j) - g_k(Tx_j)| \\
&+ |g_k(Tx_j) - g_k(Tx)| \\
&\leq \|g\|.\|Tx - Tx_j\| + \frac{\varepsilon}{3} + \|g\|.\|Tx_j - Tx\| \\
&< C.\frac{\varepsilon}{3C} + \frac{\varepsilon}{3} + C.\frac{\varepsilon}{3C} = \varepsilon.
\end{aligned}$$

Therefore

$$
\begin{aligned}
\|T^\sigma g - T^\sigma g_k\|_\sigma &= \sup\{\mathrm{r}((T^\sigma g - T^\sigma g_k)(x)) : \mathrm{r}(x) \le 1\} \\
&= \sup\{|(T^\sigma g - T^\sigma g_k)(x)| : \mathrm{r}(x) \le 1\} \\
&= \sup\{|g(Tx) - g_k(Tx)| : \mathrm{r}(x) \le 1\} \le \varepsilon.
\end{aligned}
$$

So for each $g \in B$ there exists $k \in \{1, \ldots, n\}$ such that $\|T^\sigma g - T^\sigma g_k\|_\sigma \le \varepsilon$. It follows that $T^\sigma(B)$ is totally bounded. Q.E.D.

We close with an open question:

Question 3.5. Let T^σ, the spectral adjoint of a linear mapping T between spectrally normed spaces, be spectrally compact. Is it true that T is also spectrally compact?

References

[1] B. Aupetit. *A primer on spectral theory*. Universitext. Springer-Verlag, New York, 1991.

[2] M. Mathieu. Spectrally bounded operators on simple C^*-algebras. *Proceedings of the American Mathematical Society*, 132(2):443–446, 2004.

[3] M. Mathieu and G. J. Schick. First results on spectrally bounded operators. *Studia Mathematica*, 152(2):187–199, 2002.

[4] M. Mathieu and G. J. Schick. Spectrally bounded operators from von Neumann algebras. *Journal of Operator Theory*, 49(2):285–293, 2003.

www.ingramcontent.com/pod-product-compliance
Lightning Source LLC
Chambersburg PA
CBHW060459200326
41520CB00017B/4845